Ferdinand von Mueller

Report of the Government Botanist and Director of the

Botanic Garden

Ferdinand von Mueller

Report of the Government Botanist and Director of the Botanic Garden

ISBN/EAN: 9783744650649

Printed in Europe, USA, Canada, Australia, Japan

Cover: Foto ©berggeist007 / pixelio.de

More available books at **www.hansebooks.com**

1869.

VICTORIA.

REPORT

OF THE

GOVERNMENT BOTANIST AND DIRECTOR OF THE BOTANIC GARDEN.

PRESENTED TO BOTH HOUSES OF PARLIAMENT BY HIS EXCELLENCY'S COMMAND.

By Authority:
JOHN FERRES, GOVERNMENT PRINTER, MELBOURNE.

APPROXIMATE COST OF REPORT.

DETAILED PARTICULARS.		AMOUNT.
		£ s. d.
Cost of Preparation—Not given.		
Printing (850 copies)		21 14 0

PLAN
OF THE
GOVERNMENT HOUSE RESERVE
BOTANIC GARDEN AND ITS DOMAIN
INDICATING THE PRINCIPAL PLANTATIONS

REPORT.

Botanic Garden, Melbourne,

14th September, 1868.

Sir,

In compliance with your request, I have the honor of transmitting to you a succinct general Report on the work more recently performed in the Botanic Garden and its scientific institutions.

Simultaneously, I beg to point out what measures of progressive improvements might most advantageously occupy the attention of the establishment during the next year.

Since the great excavations at the Garden lake, and the earthworks connected therewith, were completed, it became possible, within the means available, to finish the various lines of walks, which now extend in the aggregate over 22½ miles. All of these are lined with trees, unless they pass along special garden land.

A considerable extent of these walks requires, however, yet to be somewhat raised and to be covered with a gravel-layer, or perhaps with clayey grit, which is far more accessible, and will bind into a firm mass, impervious to rain. A large portion of the main drive from the City bridge to Anderson-street, needs yet to be macadamised, and basalt boulders might be used to mark off lastingly its footpaths.

The tree lines along the walks amount altogether now to 21 miles; also, different kinds of trees have recently been chosen for these avenues, to exhibit the relative merits of each. The remaining portion of the reserve between the City bridge and the Botanic Garden has latterly also been planted with many additional kinds of Pines—not less than 21,000 Pines, representing very many species, being now grouped or scattered on the lawns. To prevent more completely a certain degree of monotony, which might be caused by the massive upgrowth of conifers, though many are of very distinct form, and though lines of deciduous trees dissect the lawns, I introduced into the incipient pinetum several hundreds of New Zealand Palm-lilies (Cordyline Australis and Cord. indivisa), and also numerous groups of real Palms—for instance, the Gippsland Fan-Palm, the New Zealand Nika-Nika, the Date, the Seaforthia, the Sabal, and a few others equally hardy. Many of these Palms or palm-like plants have become already very conspicuous, and it may be readily foreseen that, within a few years, the environs of the city will assume by this measure an aspect so exotic, that a visitor viewing the suburban landscape will imagine himself to be within the tropics. To the Palm groves require still to be added in quantity the Chilian Jubæa and the equally hardy Chinese Livistonia. The various trees will form a nucleus for forest culture when gradually bearing seeds, and when not merely the protection but also the enrichment of the native forests will become an object of legislative enactments. The total number of trees now planted out approaches to 30,000. The Willow plantations along both the Yarra banks, from Prince's bridge to Richmond, have been renewed or completed this year on the municipal side of the river by the aid of the Corporation. The renewal of the fences since the last floods, effected at great expense by the City Council, has afforded for this purpose all the necessary security. Weeping Willows and various kinds of Basket Willows have been chosen promiscuously to combine ornament with utility.

A dense belt of vegetation will thus guard against accidents, embellish the river, consolidate the banks, afford more shade, shelter the Garden against the piercing westerly winds, and replace permanently the fences, apt to be carried away by the floods.

Tall Danubian Reeds, Callas, patches of Tea-tree (Melaleuca ericifolia, transferable in an upgrown state), Poplars, Ashes, Elms, Oaks, all of various kinds, Toi-Toi, Pampas Grass, Tamarix, Ampelodesmos, Wiry Muehlenbeckia, Poa ramigera, will ere long impress on the once dismal swamps and river banks a smiling feature.

The many thousand large plants required for this purpose were partly supplied by donations or interchanges. Clover and lucerne are also established on the lagoons and even on the rises.

To render, in our zone of evergreen vegetation, the Yarra valley never of winterly, leafless aspect, the City Council very kindly allowed a strip of ground all along the northern banks to be ploughed for the reception of seeds of such quick-growing evergreen trees (chiefly Eucalypts, Acacias, Exocarpus, and Casuarinas) as will resist those occasional inundations, to which we are still likely to be exposed, unless many more of the ledges of rocks across the Yarra are blasted away, to decrease still further the niveau of the river, a measure which the still rapid fall during floods will admit of.

To secure the lower part of the Garden against such calamities and destructions as were experienced during the last four floods, it will be necessary to raise the river bank still three to four feet higher, perhaps with the formation of a terrace, although the embankment has been heightened already all along the Garden to the extent of several feet. This security could, however, not be afforded on the expansive flat next to the City bridge without serious impediment to the flood stream ; but the swampy ground, now with the change of seasons wet and dry, will absolutely need deepening in several places, and raising (under formation of islands and such like ornamentation) in other spots, inasmuch as localities on which the area of dry land and of ponds is not properly defined, are prone to originate, by algic growth, malarian fevers. Consequently, on grounds of sanitary necessity alone, I feel bound to recommend this measure.

A spacious sluice was built, by Garden labor, last year, to admit of the sudden filling of the Garden lake whenever the river rapidly rises, in order that the demolition of the embankments of the lake may in future be obviated.

The tall Indian Bamboo has been acclimatized, and is, with other Bambusaceæ and the Nile Papyrus, chosen to fringe the lake. In a climate like ours, which admits of the culture of so many tropical plants without glass protection, it is always an important object to group the greatest possible number of prominently remarkable plants from various parts of the globe suitably together. This, indeed, is one of the greatest charms in our horticulture. Throughout the Garden ground numerous new species have been added annually, predominance being given to such shrubs and perennial plants as entail the least attention for maintenance. Were it otherwise, so extensive an area could not be maintained in sightliness, whilst here throughout the year the growth of weeds, annually more diversified, is to be coped with. And even now it is unavoidable to cover the central portions of all the shrubberies densely with perennial grasses, an operation which could not have been effected a few years ago, because the plants, then small, would have become suffocated. Plantations have also been formed at the stately girder-bridge, a structure which reflects high credit on the Department of Public Works. Whenever the lower part of Anderson-street is to be filled up, then the dyke now forming the approach to the bridge ought to be reduced.

The whole area of the Garden and arboreta now laid out comprises nearly 400 acres, including the lake with its six islands. To the latter, a seventh requires to be added, on the north-western extremity. By the extensive excavations on the lagoon, the once inundated eastern tea-tree ground has now been completely reclaimed, and forms a miniature forest, readily accessible to pic-nic parties from the river. Turf soil is by these means also easily obtained for Nursery culture. The work connected with the excavations also enabled me to establish passages across three of the bends of the lake, whereby the distance from point to point has been conveniently lessened. It allowed, also, widening the causeway and securing good soil for the Garden. Unrestricted access for carriages is given to all the rising ground in the reserve, from which such panoramic views may be enjoyed over the city, suburban landscapes, and bay ; and it is anticipated that, whilst from year to year the park-trees will afford augmented shade and shelter, the locality indicated will become to residents of the city one of the easiest and most favorite resorts for recreative enjoyments. A proposition, suggested in one of my former Reports, that the paths along the base of the ridges and along the Yarra banks might be widened into pleasure drives, could now be readily carried out, the Yarra flats, by recent arrangements, being no longer occupied as pasture ground.

In special artistic ornamentation as yet little has been effected, the Director deeming it of pre-eminent importance to devote his early means to the raising of trees and utilitarian plants, such as will mitigate the heat of our summer clime, and increase the salubrity of the city, or such as will play an important element hereafter in our rural economy, and originate new industries. This is the reason why no fountains exist, save one in the central island of the lake ; thus neither are statues erected.

5

Works of art we can call forth at pleasure, while time lost in forming the plantations cannot be regained. Now, however, since the main planting operations have been effected, it is but too desirable that a few appropriate statues and monumental works should add to the embellishment of the very varied vegetation, and stand with it in bold or beautifying contrast. It is proposed to gather works of art, constructed of the most varied material; the Carrara marble, all the cement compositions, the various blendings of ore, might all be brought together for illustration. For the play of fountains, the water pressure was hitherto quite insufficient, inasmuch as the Yan Yean works are only utilized when, at late night hours, the pressure exceeds 40 lbs. to the square inch. Had not, providently, each of the many Garden buildings been supplied with a spacious cistern, it would have been impossible to save the plantations from destruction during the trials of the summer months, unless by costly means Yarra water had been forced to the culmination of the hill for extensive irrigation. A special vote, adequate for such waterworks, has never been at my disposal, nor could such independent water supply have been maintained, unless annually a considerable outlay for fuel and attendance to an engine were incurred, or, what appears still less desirable, a windmill—apt to interfere with the traffic, and never sightly—had been established on the summit of the ridge. Nevertheless, it might be highly instructive to show, by local experiment, how much Yarra water could be forced by steam-power to the summits of our rises, within certain expenditure of capital and labor, because the fertility of many extensive tracts of the country could be very much increased, and the clime vastly be ameliorated, if rivers like the Yarra, and still more so those of the great Murray system, were not allowed to flow unutilized into the ocean.

Waterholes are sunk into tenacious clay soil on the higher-lying parts of the ridge, for securing the storage of Yan Yean water during rainy nights; and from these reservoirs the water is led readily during the hottest weather, by gravitation, to the plantations on the slopes below.

The abandoned quarries have been decorated with Agaves, Aloes, Mesembryanthema, some Pelargonia and other rock plants; while Brambles, Strawberries, and other wild fruit plants, attractive to children, have been planted in the gullies. Goodenias, Roses, and other shrubs line the river and lagoons. The fern-tree gully has been extended, and to the various hardy arborescent Ferns, some perhaps a century old, huge square Todeas of great age, Staghorn Ferns, and very many other species, became added in masses. The kinds of hedges now shown in different parts of the ground are very various, but that of Pittosporum eugenioides, first adopted by myself, is most admired, and called forth an extensive trade in this plant. Four other New Zealand Pittospora, as well as our native P. undulatum, are among those chosen for hedges. The Chamomile edgings, as time absorbing as defertilising and apt to be trodden down, become gradually abolished. Turkish Box, dwarf Roses, Veronica decussata, Rosemary, and most particularly Mesembryanthemum tegens, are substituted. The latter plant can be obtained largely from the Yarra flat, never fails in the heat of summer, and grows so depressed as to need only lateral trimming. Although large improvements have taken place on all the lawns, they still require gradually to be turfed with Cynodon Dactylon, a grass which is within a few weeks established, by casting its rhizoms, converted into small pieces, over the broken and levelled ground, a process extensively adopted by the Director of the Sydney Botanical Garden. It tends also much to subdue weeds. On the even surfaces of ground clothed with Cynodon, an ever verdant fine turf can be maintained by the ready appliance of lawn-cutters and rollers. Banded flower masses might be interwoven; but as yet such works of luxury, for which, after the lapse of the season, no permanent return can be shown, have not been attempted in this young establishment. There is, nevertheless, a gay display of flowers in the special garden land through the greater part of the year; indeed, the variety is far greater than a superficial observer will imagine, inasmuch as the area variedly studded with flowers is so extremely extensive.

The incessant calls, however, to provide for public fêtes, tea meetings, and bazaars, decorative flowers, not rarely deprive the Garden of a real show of ornament. The plants throughout the ground are very extensively labelled, about 3,000 iron labels being employed. Labels, however, with fused, and thus unobliterable letters, are here, as elsewhere, yet a great desideratum. In the large conservatory all plants are placed, for instruction's sake, along both sides of the stages, so as to represent those of the Western and of the Eastern hemisphere separately, the plants of the various families being again grouped together. In an inexpensive structure, far too modest to do justice to so grand a plant, the Royal Water-lily has flowered throughout two seasons, and repeatedly has ripened seeds, available for transmission to the hotter parts of Australia. The high temperature of the Victoria-house is inexpensively provided by its connection with one of the forcing pits, while, in the humid heat, Vanilla and many other epiphytal Orchids of the jungles of the torrid zone find here the conditions necessary for their permanent existence. The standard collection of Vines and Orchard trees has annually been added to. Fruit from these has been supplied to public charities. The experimental ground has also annually grown richer. To attempt to specify the treasures of the Garden, whether utilitarian or ornamental (many first introduced by the Director into Australia), is beyond the scope of these pages. The special catalogue appended to this document will exhibit many which we possess, but not all, inasmuch as thousands of plants occur yet in too young a state to correct their erroneous appellations. Mere varieties and garden hybrids, as a general rule, have been excluded from the catalogue. In a full account of the botanical establishment, submitted by order of the Government to His Royal Highness the Duke of Edinburgh, I specially alluded to some of the leading useful or remarkable plants. But an explanatory enumeration of all would enlarge to a volume, or might find space in a contemplated publication, which would serve as a garden guide. To add still further to this valuable collection, Mr. Heyne proceeded, at my request, early this year to Sydney, to select from the local conservatories. In this object he was very liberally supported by Mr. C. Moore. The suppression of the two principal kinds of Mistletoe (Loranthus pendulus and Loranth. celastroides), which, on neglected ground, often manifest themselves by the widely visible dead ramifications of the trees, causes here much loss of labor. The annihilation of the trophy guns throughout Britain suggests the propriety of removing those which occupied for some years a position in this Garden. The spot allotted to them might far more pleasingly be occupied by a small ornamental building, in which the native birds, which, permanently or migratively, are inmates of the Garden area (approximately 140 species), could be illustrated by single museum specimens, to satisfy constant inquiry in reference to the scientific names of the species. The lake is often swarming with water birds, the tame swans, pelicans, ducks, &c., acting as decoy birds. Thrushes teem in the shrubberies. To the aviary, donations of parrots, cockatoos, and other showy native birds, not formerly kept, would add much interest. The formation of an outdoor fresh-water tank, for the culture of hardy aquatics, which in the lake generally succumb under the prey of water birds, is highly recommendable. The introduction and multiplication of important plants, of industrial or medicinal value, has received careful attention. Thus, about 10,000 young Peru bark plants have been raised, comprising mainly Cinchona succirubra, C. calisaya and C. officinalis, the latter, the most hardy of all, predominating.

These plants have withstood the night frosts, which we experience near Melbourne, when merely placed in brush shades. On one occasion the thermometer in these shades sunk to 28° F., while in the open ground it stood at 24° F. near the surface; still the plants suffered not further than getting some of the leaves and youngest branches injured, but soon formed new leaf-buds. These frosts affect, moreover, also some of the plants which inhabit the mild sheltered glens of our ranges, and I am, therefore, justified in anticipating that, in many of the warmer forest regions of Victoria, the Cinchonae could be grown to advantage, these plants being consociated with Fern trees in their native haunts in the middle regions of the Andes. Coffee plants scarcely suffered in the brush shades, in which the temperature may be regarded almost analogous to that of our fern-tree gullies. It would be very important to ascertain, by actual test in the ranges, whether the Coffee and Cinchonae would yield prolifically. In such localities, under any circumstances, the Tea-shrub would so luxuriate as to produce an abundant crop of leaves, since even in dry localities of the Botanic Garden, and in its poor soil, the Tea-bushes have grown quite well. Cork-Oaks, of which, like of Tea, several thousand plants are reared, would also produce far more rapidly their useful bark in the ranges than near the city; there the American Hickories and Walnuts, of which a copious supply of seedlings exist, would grow much faster. These, with the Red Cedar, West Australian Mahogany, Sumach, Scotino, Dates, Carob trees, Valonia and Dye-Oaks, Mastix trees, Arrowroot, and perhaps also Tapioca, Tamarinds, and very many other prominently utilitarian plants, would thrive best in the rich humid soil of our mountains, and might occupy localities not readily eligible for cereals.

Observations in reference to the effect of night frosts on the principal plants, as well as records concerning the flowering time of various species, are registered in the office. Notes are also accumulating respecting the adaptability of the dry desert tracts, and again of the alpine highlands, to certain cultures. By a Parliamentary return submitted last year, it was shown that, from 1859 till July 8, 1867, not less than 355,218 plants were distributed to the public reserves, cemeteries, church and school grounds of Victoria. During 1868, again, 49,475 plants were rendered available for this purpose. These distributions comprised very many of the rarest Pines and other select plants, often not otherwise available, many requiring two years' attention in the nurseries here, thus involving the necessity of maintaining, during some years, approximately, 40,000 plants alone under pot culture. Bearing in mind the increasing extent of trading establishments, exceedingly well conducted, it is worthy of the consideration of the Government whether these distributions from a public establishment should not be materially decreased, or abolished altogether. Numerous plantations, by the impetus given, are now established on public grounds throughout the colony, from whence, moreover, seeds and cuttings might be locally obtained. Eminently useful plants of many kinds have, for local experiments, been widely scattered over the country. The Treasury Reserve received last year 245 to some extent already upgrown coniferous trees.

Turning to the special phytographic department, it may be observed, that the Museum now contains about 350,000 prepared and arranged plants ; the Australian portion being richer than that of any kindred institution in existence.

The sixth volume of the *Fragmenta Phytographiæ Australiæ*, a work devoted to original discoveries, and written in a language common to science of all nations, is almost completed. The fourth volume of the universal work on the plants of Australia is, through my aid, under the rare advantages attainable in the great national institution of Kew, just completed by the President of the Linnean Society, and comprises the orders of Corolliflorae. Extensive preliminary researches have been carried on already for the fifth, sixth, and seventh volumes; to which, finally, a supplement is to be added. To promote, by further field researches, the objects of this large work, on which all subsequent medical technological and rural observations in reference to the native Australian vegetation must rest, I visited, during the past spring, one of the most important tracts of West Australia. Finally, also, the great task yet remains to be performed of tracing out more completely the relation of geology to the distribution of the plants existing as well as passed away—a line of researches for which excellent geographical and geological maps are annually affording more facilities. Mr. Dallachy continues sedulously to collect, both for the Garden and the Phytographic Museum, in the north-east part of Queensland.

The following are the genera which, since the issue of my last Report, by local independent researches, have been added to the system of Australian plants:—Dillenia, Cakile, Aldrovanda, Gomphandra, Connarus, Strongylodon, Salacia, Caryospermum, Casearia, Cucurbita, Œnanthe, Antirrhœa, Lasianthus, Ophiorrhiza, Geophila, Aniseia, Erycibe, Ichnocarpus, Ceropegia, Bassia, Chrysophyllum, Thuubergia, Graptophyllum, Dischisma, Cylicodaphne, Cinnamomum, Plecospermum, Taxotrophis, Hyrtanandra, Nepenthes, Apostasia, Cirropetalum, Pogonia, Spathoglottis, Dracæna, Bambusa, Centotheca, Angiopteris, Marattia, Deparia, Isoetes ; and the following genera, new to phytography:—Fitzgeraldia, Pagetia, Davidsonia, Thespidium, Eleutheranthes, Thozetia, Carnarvonia, Darlingia, Helmholtzia, Corynotheca ; by which means representatives of Connareæ, Samydeæ, Selagineæ, Nepentheæ, and Apostasiaceæ are added to the Australian flora. The following are additions to the list of Australian trees published in the volume of the Intercolonial Exhibition:—Melodorum Maccreai, Pittosporum rubiginosum, P. venulosum, Eriostemon squameus, Sterculia laurifolia, Sloanea Woollsii, S. Macbridei, Gomphandra Australiana, Leucocarpon celastroides, Taxitrophis rectinervis, Ficus Benjaminea, Croton triacros, Beyera viscosa, Mallotus polyadenos, M. Dallachyi, M. repandus, M. Chinensis, M. pycnostachys, Macaranga involucrata, Oxylobium Callistachys, Pithecolobium Sutherlandi, Archidendron Lucyi, Quintinia Fawkneri, Cuttsia viburnea, Hakea macrocarpa, Carnarvonia aralifolia, Dryandra floribunda, Myrsine achradifolia, Bassia galactodendron, Chrysophyllum pruniferum, C. myrsinodendron, Alstonia verticillosa, A. villosa, A. excelsa, Cerbera Odollam, Casuarina Fraseriana.

In the event of its proving inadvisable to devote the New Exhibition-building to the intended collections of a general industrial museum, it might be desirable to enlarge the Phytological Museum-building in the Garden, in order that a full display of vegetable objects of industrial interest may be formed. The absolute want both of space and accommodation frustrated every attempt to render my establishment also useful in this direction.

During the Intercolonial Exhibition an apt opportunity arose to represent more fully the technological value of many native vegetable products, and for this purpose, from the ordinary resources of the establishment, a laboratory was constructed. I need not detail the experiments conducted in reference to the value and percentage of many kinds of paper material, essential oils, dye stuffs, wood vinegar, tar, wood spirits, and tannic acid, from native plants, especially trees ; on all of which ample information was offered in the documents concerning the Exhibition. These phyto-chemical observations have since been continued as far as circumstances permitted.

Appended to this Report are the tables of very extensive series of analyses, conducted in detail by Mr. Chr. Hoffmann, in reference to the yield of potash in our more gregarious native trees. They show that the manufacture of this alkali can be pursued here more profitably than in those countries in which the supply of original timber is far less extensive than in Victoria. The examination into the yield of iodine and bromine in our seaweeds is commenced ; likewise, the yield of soda in one of the principal littoral plants is recorded. I have entered also on a series of toxicological researches, by which it is hoped the nature of those poison plants so injurious to stock will be fully elucidated.

A supplementary catalogue of the library is also given ; many of these works, however, had to be provided by the Director's private means.

It yet remains for me to record my sense of obligation to the very numerous donors, who enriched the various branches of the establishment during a more recent period. A glance at the list of these supporters will also be the most convincing proof of the wide external communications of the department, while a reference to the plan annexed will at once largely explain the extent of the internal operations, which are singularly multifarious. It would be unjust were I not specially to allude to the graceful concession continued by the Peninsular and Oriental Steam Navigation Company, the owners of the *Great Britain*, and many other mercantile and seafaring gentlemen, to convey, freight free, the consignments of this establishment, or were I to pass silently the kind aid rendered by His Excellency Sir Henry Barkly, in effecting from Mauritius the final transits to Bourbon and various parts of South Africa. The foreign communications involve the necessity of correspondence in several languages, the number of all letters issued being about 3,000 a year. The permanent property in buildings, iron fences, drains, boulders, waterworks, collections, library, and lasting improvements, irrespective of the plants distributed, and irrespective of the value of the local plantations, fell not short of £27,000, according to an estimate made two years ago by professional gentlemen not connected with the department. This lasting property increased, by additions since, considerably in value. Nor is in this estimate the value of the iron bridge included. While aiming, as far as in his power, at the utmost economy, the Director hopes that those means which Parliament may also in future be pleased to entrust to him will proportionately enhance the lasting value of the establishment, and bear, in scientific information afforded, and in practical services rendered, always an ample return.

I have the honor to be, Sir,

Your very obedient servant,

FERD. VON MUELLER, M.D., F.R.S.,

Government Botanist and Director of the Botanic Gardens.

The Honorable J. M. Grant, M.P.,
  President of the Board of Land and Works,
    &c., &c., &c.

SUPPLEMENTARY REPORT.

Melbourne Botanic Garden,
8th March, 1869.

SIR,

In accordance with your instructions, I have the honor of submitting a brief Report on the work carried on in the Botanic Garden, and the scientific establishments connected therewith, during the last six months. This document may be considered as supplementary to the last general Report, and will also briefly explain what additional work seems recommendable during the year 1869.

In the horticultural division of the establishment, the shelter accommodation for tender or young plants has been extended so much, that now the whole space under cover, either by glass or calico or brush shades, exceeds half an acre. Many rare plants, often new to Australian cultivation, flowered or bore fruit for the first time. To show how the riches of the establishment are thus yearly increasing and may extensively be diffused, I may instance that the first Flame-tree, in producing fruit last year, gave the means of raising nearly one thousand seedlings. The Grevillea avenues commenced flowering this season, and it may be imagined, what a brilliant effect the long lines of this tree will produce in years to come.

The conservatories have been rendered lately still more gay by new access to the silvery and banded Assam Begonias, the variedly spotted Caladiums of Central America, and various gesneriaceous and many other gorgeous plants; while arrangements are made to add to the collection Dionæa, the Sarracenias of North America, Biophytum, and other plants, remarkable for spontaneous movement or extraordinary structure. The Great Central American Water-lily bearing the name of Her Majesty is now flowering through the third year; but the narrow, inexpensive house, allotted as well to this noble plant and other tropical aquatics as to the equinoctial Orchidæ, stands much in need of extension. To the plants in the general garden ground additions have steadily been made, so much so, that now a fair re-arrangement can be effected in many places, to represent on separate plots the characteristic vegetation of the great divisions of the globe in a very instructive manner. During the extraordinary dryness of this summer miles of edgings became quite parched, and will require renewal in the autumn, for which purpose the less perishable Mesembryanthemum will be chosen. Porcelain labels, with unobliterable letters, have been ordered as a commencement of naming the plants in a more lasting and sightly manner. His Royal Highness Prince Alfred, during his stay last year, condescended to plant on one of the lawns, in remembrance of his visit, the Patagonian Saxono-Gothæa conspicua and the Californian Pinus Alberti, trees which commemorate the name of his illustrious and lamented parent.

A great boon has been conferred on the Garden by the Government, in sanctioning the establishment of steam works for forcing Yarra water to the highest rise, 110 feet, from whence some irrigation is now effected over the greater part of the Garden area and the adjoining reserves. If even during ordinary summers the duty of providing for the safety of the extensive plantations proved a source of very great anxiety, and of extreme toil, both day and night, then this duty became still more onerous during the horrors of an almost rainless summer, when, during successive hot winds, the up-growing tree-vegetation, as well as the tender garden plants, had to be protected over nearly 400 acres of ground against the imminent danger of destruction, and this with an inadequate water supply. Happily this difficulty, in a great measure, has now been overcome; and inasmuch as it may be of importance to owners of estates on the sides of rivers to obtain data of the yield and working expenses of the engine

No. 21, a.

employed at this spot (which engine is one of high-pressure and six-horse power), I beg to submit the following calculations :—

Yield—In one hour	...	...	...	...	...	4,962	gallons
In one day, under actual working, of ten hours			...		49,620	,,
In one year, with 308 working days of ten hours' work		...		15,282,960	,,

				£	s.	d.
Expenditure—Wages of an engine-driver for eight hours each day, at		...	0	8	0
Wages of a mechanic for two hours each day, at ...		...	0	2	0
Four and a half cwt. of coals each day...	...	...	0	5	11
Oil, nearly 1 lb. ...	...	...	...	...	0	0	7
Tallow, cotton-waste, and anti-friction grease	...	...	0	0	10
Daily expenses ...	...	...	...	£0	17	4

Expenditure per year of 308 working days, 17s. 4d. per day ... £266 18 8

The expenditure for the raised Yarra water exceeds thus, slightly, 4d. per 1,000 gallons.

New South Wales house coal, screened, per ton of 2,240 lbs., as under contract for 1869, £1 6s. I cannot state the precise value of the engine, it being transferred from another department.

The above calculation allows, however, not for occasional repairs, nor for a few hours' detention of the work during each week for cleaning purposes, nor for interest on capital expended for the engine, force-pump, and water-pipes.

The capacity of the small temporary tank to receive the water at the summit of the ridge is, however, only 1,700 gallons, and until provision shall have been made for a spacious and raised tank, as intended, one great difficulty will continue, namely, that although a large supply of water is available it can, under faint pressure, only in very limited quantity find its way through the ramifications of the former Yan Yean pipes to distant higher parts of the Garden and reserves.

The eight mostly spacious cisterns for the reception of rain-water from the roofs of the Garden buildings, and the four iron tanks, will be kept filled, to provide against any emergency in the event of breakage at the engine. I may still remark that, although during the cooler months steam-power will not require to be used every day, nevertheless, any savings then effected in the outlay will need to be expended again during the hottest weather, when fourteen hours' daily work of the engine will be needed.

The Geyser fountain in the lake (which for two afternoon hours in cool weather, and then on Sundays only, was worked with Yan Yean pressure) has ceased to play. Until the steam-engine was provided the Garden enjoyed Yan Yean supply during two night hours (from 3-5 a.m.), provided in cool weather the pressure admitted of obtaining any supply at all; but this boon has now entirely ceased. The whole of the former Yan Yean pipes, provided on expenses of the Garden, have become available again for the conveyance of the Yarra water.

The large reserve between the St. Kilda road and the Yarra is converted, within the last five years, from a treeless waste into an incipient forest. From year to year additional kinds of trees become interspersed; thus shade and shelter as well against the north-western desert winds, as also against the south-west antarctic storms, will be more and more obtained. Few even of our metropolitans seem aware that the verdant valleys which, within five minutes' drive from the City bridge, slope gently to the Yarra, afford already charming picnic grounds, on which, free from the dangerous vicinity of the reptiles of our ranges, field amusements can be enjoyed simultaneously with views of rare beauty. Access of carriages to the whole of this rising ground and its gullies is permitted, under the anticipation that all ordinary caution will be exercised to prevent injury to the young trees. By the gradually denser growth of grass, lucerne, and clover plants, the so-called Cape-weed (Cryptostemma calendulaceum) has become largely suppressed; but inasmuch as the Director of the grounds has repeatedly been accused of having brought this and other weeds, as well as some winged invaders, into our colony, it may be right to place it here on record, that the whole of these assertions is contrary to facts, and that already, in 1833, Baron Von Huegel noticed and recorded the Cryptostemma as an inexterminable weed of Australia. A gardener's cottage occupies, since a few months, the last of the empty old quarries, until then a favorite retreat of vagrants.

For more than a mile's length, basalt boulders have recently been brought from Jolimont, by permission of the City Council, to line the intended footpaths on both sides of the main drive. The drive itself, to the width of twenty feet, requires to be macadamized, for which purpose the boulders may be utilized, whenever more elegant linings can be substituted for them.

By the friendly aid of the military authorities lately, walks became laid out on and near the Yarra bank, towards the City bridge. During the coming autumn it is intended to define these walks with many hundreds of rose-bushes. The fences along the St. Kilda road, Domain road, and Anderson street, up to the point at which the iron fencings commence, have sunk almost into destruction. Several thousand young Willows, planted along both sides of the Yarra bank during the last cool season, have weathered fairly through this summer of drought, labor for watering those on the north bank having been granted by the Corporation.

An important work will devolve on the department in further excavations on the lake, if the needful extra aid can be rendered. The water evaporated entirely through the aridity of the season, and no sufficient rise of the river has taken place to refill the lake. The advantages of deepening this basin would be manifold. Its niveau and that of the river would become permanently equal, and a constant communication between both would become possible; material would be gained to heighten the flood-dam so far as to obviate future inundations of the Garden; the brackish water of the lake would become fresh and available for garden purposes; further storage of soil for the improvement of the meagre Garden slopes would become possible; waterfowl might permanently be maintained on the lake; and finally, the aspect of the whole landscape would be greatly beautified.

Sir William Macarthur's method of wrapping hard seeds into moistened cloth to speed their germination has been adopted to advantage.

A variety of Bamboos and different Sugar-canes were secured, including the hardy Chinese cane; forty-eight kinds of Vines were added on behalf of the Acclimatisation Society to the already large collection, which includes the white and black American Scuppernong, the Sultana raisin grape, the French Cognac grape, Follet Blanche, and many other famed kinds, new or rare in Australia. The true Oriental Dye Saffron, Colchicum, the oil-yielding Sesamum, the Tussac grass of the Falkland Islands, the Caper (quite an ornamental plant), the wide-spreading avenue Acacia of West Australia (Acacia saligna), Ficus sycamorus (the best of all avenue trees of the Orient), the Clove, Rhamnus utilis (yielding the green satin dye of Chim), the Sapodilla, the Avocado Pear, the Indian Teak, Cassava, Squill, Turmeric, the medicinal Bhel fruit, the Tree Cotton, Mangostan, edible Vanguiera, Aya-pana, Gelsemium, and many other important plants, are more recent acquisitions to the garden. Although it may as yet be impossible to cultivate remuneratively the Saffron and many other of the plants indicated, it remains evidently still the aim of a public institution to establish such plants timely in the country.

Turning to the nursery department I can report favorable progress, notwithstanding the precarious supply of water during the great heat. For the first time in Australia masses were raised of plants of Assam Tea (the seed kindly supplied, on the Director's wish, by W. H. Birchall, Esq.); so also large numbers of the White-heart Hickory or Mocker-nut (Carya tomentosa), of the delicious Pecan-nut (Carya oliviformis), the Butter-nut (Juglans cinerea), the Black Walnut (Juglans nigra), the Himalayan Oak (Quercus incana), the Chesnut Oak (Querc. Castanea), the American Swamp Oak (Querc. Prinos), the Bur Oak (Q. macrocarpa), the White Oak (Q. alba, a most valuable timber tree), the Jersey Pine (Pinus inops), the American Pitch Fir (P. rigida), the Douglass Pine, the noble Himalayan Pinus longifolia, the Chinese Fir, the Balm of Gilead Fir (P. balsamea), the double Canada Balsam Fir (P. Fraseri), the West India Pencil Cedar (Juniperus Bermudiana), and the American Cherry Birch (Betula lenta).

Many other highly valuable trees have been lately introduced, but not really in masses. Secured were, however, large supplies of the seeds of Pinus Gerardiana (the Tibet Ree or Shungtee), which furnishes sweet edible nuts for Indian and Persian bazaars; and grains also were obtained in quantity of Juniperus religiosa (the Himalayan Pencil Cedar). Many good-sized plants of the latter are since several years on our lawns. Nearly all the tree seeds from the United States were obtained through the generous aid of Prof. Asa Gray, of Boston.

Perhaps the most remarkable of all plants lately brought under cultivation is the deadly poisonous Physostigma venenosum, the Calabar Ordeal Bean, a plant of the utmost importance in ophthalmic diseases. The large hard bean was buried fully four years in soil before it germinated.

As decennia roll on, many of the trees, which under great efforts are now introduced, will undoubtedly bear prominence in our forest culture, a great subject which more and more presses on legislative attention, since already so much of the native timber in all the lowlands has been consigned to destruction. If, in densely populated countries like Belgium, one-fifth of the whole of its territory is scrupulously kept under forest culture, it ought to be a final aim, in a far hotter clime, to maintain a still greater proportion of its area covered by woods, if the comforts and multifarious wants of a dense population are to be timely provided for. It is especially in the western and northern parts of Victoria where exertions in this direction have to be made ; it is there where extensive shelter and retention of humidity is needed, and there also where artesian borings on spots, indicative as eligible, would vastly promote the raising of forests.

By your kind concession, Sir, I was enabled to spend in the beginning of this year one week in Tasmania with a view of adding, by field observations and new collections, to the material of my works. This journey (my first to the island) was to me replete with interest. For although I had aided in the elucidation of the Tasmanian vegetation for more than twenty years from museum plants, I had no opportunity until this year to observe the many highland plants, absolutely peculiar to the island, in their wild native grace. Moreover, I succeeded, within the brief time of my visit, to ascend Mount Field East, about 5,000 feet high, lying about half way between Hobarton and Macquarie Harbor; and as this mountain range and the shores of Lake Fenton had not been subjected to any previous phytological investigation, it fell to my share to obtain copious novel information on the distribution of the alpine plants of Tasmania. To contrast the consociations of these and their geological relations with those of the Australian Alps proved in a high degree instructive.

The Museum collections become more and more important, and their value as a lasting source of authentic information for centuries to come can never be over-estimated. It remains, however, a source of regret that no more amateur collectors in far inland localities send spontaneously plants, simply pressed and dried; by which means much would be learnt additionally on the range of different species over the continent, and their variation in form. The facilities for obtaining on any plants reliable information, always cheerfully given, might in all future also not be equally great, nor the opportunities of literary record always remain the same. If to the several hundred thousand plants in the Museum still a collection could be added, rich in authentic specimens, described in works during the earlier parts of this century, we would then possess one of the grandest institutions for phytographic research anywhere in existence.

The want of an appropriate hall, with proper fittings, has prevented special teaching by lectures in the Garden. But as an illustrious Professor of Natural Sciences also teaches phytology at the University, it might be desirable to restrict any future occasional demonstrative lectures in this place to those industrial phytological subjects, through which science enters into occupations of daily practical life, occupations of which many in this young country have still to be called forth. It might be desirable also, with a view of diffusing a vivid knowledge on the native vegetation, to arrange for occasional Saturday afternoon excursions of students and amateurs to botanically interesting spots in the vicinity of the city.

Whatever may be the decision in reference to the organization of the general Industrial Museum in the city, there should certainly one spacious room in the Garden likewise be available as a store of object of leading importance, emanating from plants of different parts of the globe. Such vegetable objects, like those in Sir Will. Hooker's great institution of Kew, could no more advantageously be studied than in connection with the living plants of the Garden or conservatories here.

The timber, fibres, resins, gums, dyes, paper-materials, drugs, oils, alkalies, and many chemical educts from plants of Australia could be contrasted with similar products of other countries; the processes of manufacture and their technological and commercial value be demonstrated; while subjects relating to culture of any kind could be elucidated, diseases of plants by objects and drawings illustrated, and many other kindred enquiries drawn into the vitality of practical application. Thus I may instance that it seems not generally known how our common Eucalyptus leaves under Ramel's process can be converted into cigars, or how the same leaves serve as a remedy in intermittent fever.

I herewith beg to submit the fourth volume of the work on all Australian Plants, elaborated, under my aid, by the President of the Linnean Society. This volume brings the number of species already described to nearly 5,000. For the fifth volume, which is to embrace mainly the Monochlamydeæ, the whole material in our Museum has been preliminarily prepared. Hitherto, precisely fifty

13

large cases of museum plants in 922 large fascicles, with notes, have been transmitted on loan to Kew for the elaboration of this work, the collections here accumulated, or furnished originally from hence, being more extensive than the united former Herbaria of Australian plants in Britain.

We may reflect, not without pride, on the fact, that a similar descriptive work exists not even yet for the vegetation of Europe, and we may also remember that, without a work of this kind, the confused vernacular appellations and any medicinal technological cultural, or other observations on the native plants, could not be reduced to a solid scientific basis. R. Brown's celebrated Prodomus, issued in 1810, comprised only about one-third of the Australian plants then known, and even the orders elaborated in his volume have been augmented by more recent researches almost threefold. Of the Fragmenta Phytographiæ Australicæ, the sixth volume has also been completed last year, and the seventh is commenced. Within the few next years I trust it will be in my power, if Providence grants me life and strength, to issue, on the plants of each of the Australian colonies, a special volume, for which much preliminary work has been done.

The library became also lately further enlarged, but mainly on the Director's private means. Personal travelling expenses since 1852, and all outlay for scientific and local journals, British and foreign agencies, means of conveyance for attending at the city, office light, and many other official expenses, as well as the courtesies which are demanded from a public department frequented by very numerous visitors, have also ever solely and readily been defrayed from the administrator's own resources, who, not for any selfish purposes whatever, ventures to place these facts, after the lapse of many years, on record, but simply in justice to himself, because the obligations devolving on him in maintaining the efficiency and dignity of the department seem not at all understood.

When now long past the zenith of ordinary life, he can with fairness assert, that thirty of his best years have been absorbed almost entirely in phytologic and cognate pursuits ; that almost seventeen years have been devoted cheerfully and exclusively to the main foundation and on struggling services of his department, and this, he may add, with the sole aim of endeavoring to effect some lasting good to the great country which, since twenty-two years, he adopted as his permanent home.

I have the honor to be, Sir,

Your very obedient and humble servant,

FERD. VON MUELLER.

The Honorable J. M. Grant, M.P.,
President of the Board of Land and Works.

ANALYSIS OF HALOCNEMUM AUSTRALE.

100 parts of the freshly-gathered plant contain:	
Water	84·09
Dry substance	15·91
	100·00

Percentage of ash in "dry substance," 30·62.

100 parts of the ash contain:		
Soluble	Chloride of sodium	70·34
	Chloride of potassium	8·13
	Carbonate of soda	3·14
	Sulphate of soda	7·11
Insoluble		88·72
		11·28
		100·00

The total amount of carbonate of soda, calculated from the chloride of sodium, sulphate of soda, and including that already present in the form of carbonate, is 72·17.

One ton of the freshly-gathered plant would, calculating from the above data, furnish about 212 lbs. of common commercial crystallized carbonate of soda.

By simple exposure in a room the freshly-gathered plant lost 75·6 per cent. of its water.

The percentage of chloride of sodium in seawater averages 2·71.

CONTENTS OF POTASH IN INDIGENOUS TREES.

TABLE 1.—LEAVES.

Species of Tree.		Moisture in a hundred parts.		Ash in a hundred parts of—		Ratio of the soluble constituents of the ash to the insoluble, centesimally expressed.		Potash in a hundred parts of—				Colour of the Ash.
						In Water.						
Systematic Name.	Vernacular Name.	Water.	Dry Substance.	Fresh Leaves.	Dry Leaves.	Soluble.	Insoluble.	Fresh Leaves.	Dry Leaves.	Ash.	The soluble portion of the Ash.	
Casuarina quadrivalvis	Drooping sheoak	55·08	44·92	1·95	4·34	35·81	64·19	0·21	0·48	11·07	30·91	Very light grey.
suberosa	Erect sheoak	51·50	48·76	1·61	3·73	44·74	55·26	0·31	0·65	17·38	38·50	Light reddish grey.
Banksia Australis	Common honeysuckle	52·31	47·69	2·53	5·30	20·50	79·41	0·21	0·44	8·26	40·12	Light grey.
Acacia decurrens	Wattle acacia	53·75	46·25	3·00	6·49	23·07	76·93	0·19	0·41	9·34	37·48	White, yellowish tinge.
Melaleuca ericifolia	Swamp tea tree	57·61	42·19	2·80	6·64	35·41	64·59	0·15	0·37	5·56	15·98	Reddish grey.
Eucalyptus globulus	Blue-gum tree	49·69	50·31	2·91	5·84	17·75	82·25	0·31	0·82	10·68	59·72	Light grey.
rostrata	Red-gum tree	48·98	51·02	3·09	6·04	22·66	77·34	0·49	0·82	13·61	60·96	Light grey.
viminalis	Manna eucalypt	55·19	44·81	3·75	4·18	47·42	52·58	0·49	1·15	27·65	58·31	Light reddish grey.
melliodora	Small-leaved box tree	50·75	49·25	2·28	4·63	29·70	66·91	0·40	0·81	17·46	43·89	White, yellowish tinge.
obliqua	Stringy-bark tree	57·62	42·38	1·50	3·68	33·09	66·91	0·21	0·50	13·67	41·31	White, yellowish tinge.

NOTE.—(a) Without exception, the leaves were in all cases detached from the branchlets.
(b.) The percentage of water in the leaves of those species of trees the issues of which are embraced by the bracket is slightly augmented by loss by essential oil. The greatest error resulting from this cause would occur in the percentage of water in the leaves of Eucalyptus globulus (which afford a larger percentage of essential oil than any of the other leaves here employed), and could here amount to 0·72 per cent.
(c.) All the material for analysis was gathered in March and April.

TABLE 2.—BRANCHLETS AND BRANCH-WOOD, WITH BARK.

Species of Tree.		Moisture in a hundred parts.		Ash in a hundred parts of—		Ratio of the soluble constituents of the Ash to the insoluble, centesimally expressed.		Potash in a hundred parts of—				Colour of the Ash.
						In Water.						
Systematic Name.	Vernacular Name.	Water.	Dry Substance.	Fresh branchlets and branch-wood with bark.	Dry branchlets and branch-wood, with bark.	Soluble.	Insoluble.	Fresh branchlets and branch-wood, with bark.	Dry branch-wood with bark.	Ash.	The soluble portion of the Ash.	
Casuarina quadrivalvis	Drooping sheoak	40·62	59·38	2·34	3·94	7·98	92·02	0·07	0·12	3·13	39·22	White, yellowish tinge.
suberosa	Erect sheoak	36·87	63·13	1·56	2·47	14·33	85·67	0·13	0·21	8·46	39·04	Light reddish grey.
Banksia Australis	Common honeysuckle	49·28	50·72	1·21	2·38	19·91	80·06	0·08	0·15	6·36	31·99	Light reddish grey.
Acacia decurrens	Wattle acacia	45·12	54·68	1·72	3·13	17·71	82·29	0·16	0·29	9·33	37·65	Very light grey.
Melaleuca ericifolia	Swamp tea tree	46·74	51·66	2·45	4·80	20·81	79·19	0·14	0·28	8·95	28·80	Brownish grey.
Eucalyptus globulus	Blue-gum tree	46·56	53·44	2·00	3·74	12·16	87·14	0·14	0·26	7·08	55·03	Light brownish grey.
rostrata	Red-gum tree	44·56	55·44	1·97	3·37	11·87	88·13	0·11	0·25	5·74	48·36	Light brownish grey.
viminalis	Manna eucalypt	47·69	52·31	0·84	1·60	20·20	79·80	0·05	0·18	11·13	55·10	Light brownish grey.
melliodora	Small-leaved box tree	39·87	60·13	2·19	3·64	11·37	88·63	0·08	0·13	3·59	21·57	White, yellowish tinge.
obliqua	Stringy-bark tree											

TABLE 3.—TRUNK-WOOD, WITHOUT BARK.

Species of Tree.		Moisture in a hundred parts.		Ash in a hundred parts of—		Ratio of the soluble constituents of the Ash to the insoluble, centesimally expressed.		Potash in a hundred parts of—				Colour of the Ash.
						In Water.						
Systematic Name.	Vernacular Name.	Water.	Dry Substance.	Fresh Trunk-wood.	Dry Trunk-wood.	Soluble.	Insoluble.	Fresh Trunk-wood.	Dry Trunk-wood.	Ash.	The soluble portion of the Ash.	
Casuarina quadrivalvis	Drooping sheoak											
suberosa	Erect sheoak				0·80	53·98	46·02		0·70	33·94	62·87	Very light reddish.
Banksia Australis	Common honeysuckle				1·06	60·11	39·89		0·29	27·69	45·91	Slate-coloured.
Acacia decurrens	Wattle acacia	38·97	61·03	0·50	0·82	16·34	83·66	0·05	0·08	9·77	59·79	White, greyish tinge.
Melaleuca ericifolia	Swamp tea tree	46·88	53·12	0·78	1·47	50·81	49·19	0·11	0·20	13·73	27·62	Very light reddish.
Eucalyptus globulus	Blue-gum tree	52·19	47·81	0·37	0·77	26·16	73·84	0·06	0·12	15·79	60·36	White, yellowish tinge.
rostrata	Red gum tree	51·25	48·75	0·18	0·37	48·76	51·24	0·05	0·10	27·98	57·18	Light brown.
viminalis	Manna eucalypt	44·56	55·44	0·41	0·70	17·27	82·73	0·08	0·09	10·60	61·38	Light yellowish.
melliodora	Small-leaved box tree	32·25	67·75	0·75	1·11	6·50	93·50	0·02	0·04	3·36	81·68	White.
obliqua	Stringy-bark tree				0·14	31·92	68·08		0·03	19·01	59·55	Light reddish.

TABLE SHOWING THE AMOUNT OF PEARL-ASH AND PURE POTASH AFFORDED BY ONE TON OF "FRESH BRANCHES, WITH LEAVES," AND ONE TON OF "FRESH" OR "DRY" TRUNK-WOOD, WITHOUT BARK.

Species of Tree.	Branches, with Leaves as freshly lopped off the tree,* would furnish—		Fresh—		Dry—	
			Trunk-wood, stripped of its bark, would furnish—			
	Soluble Salts may be regarded as equivalent to Pearl-ash.	Pure Potash.	Soluble Salts may be regarded as equivalent to Pearl-ash.	Pure Potash.	Soluble Salts may be regarded as equivalent to Pearl-ash.	Pure Potash.
Casuarina quadrivalvis	6 lbs. 1 oz.	2 lbs. 3 oz.	..	..	7 lbs. 3 oz.	4 lbs. 6 oz.
suberosa	7 „ 12 „	3 „ 12 „	..	..	14 „ 5 „	6 „ 6 „
Banksia Australis	6 „ 3 „	2 „ 1 „	..	..	2 „ 15 „	1 „ 13 „
Acacia decurrens ..	7 „ 12 „	3 „ 10 „	1 lb. 14 oz.	1 lb. 2 oz.	16 „ 13 „	4 „ 8 „
Melaleuca ericifolia	13 „ 4 „	3 „ 5 „	8 „ 14 „	2 „ 6 „	4 „ 8 „	2 „ 11 „
Eucalyptus globulus	8 „ 5 „	4 „ 12 „	2 „ 3 „	1 „ 5 „	4 „ 0 „	2 „ 4 „
rostrata	5 „ 6 „	2 „ 11 „	2 „ 0 „	1 „ 2 „	3 „ 2 „	1 „ 13 „
viminalis	5 „ 1 „	2 „ 14 „	1 „ 11 „	1 „ 1 „	1 „ 0 „	0 „ 14 „
mellidora ..	7 „ 8 „	2 „ 12 „	1 „ 2 „	0 „ 9 „	0 „ 14 „	0 „ 7 „ †
obliqua	..	..	..	..	..	..

According to Ilées, one ton of	Pine wood furnishes	..	..	..	..	..	1 lb. 0 oz.
	Beech	..	..	..	..	..	2 „ 14 „
	Ash	..	..	..	..	..	1 „ 11 „
	Oak	..	..	..	..	..	3 „ 6 „ †
	Elm	..	..	..	..	..	8 „ 12 „
	Willow	..	..	..	..	..	6 „ 6 „ †

According to Berthier, the ash from one ton of	Oak-wood furnishes	..	..	..	..	6 lbs. 12 oz.	..
	Lime	..	..	..	..	12 „ 2 „	..
	Birch	..	..	..	..	3 „ 9 „	..
	Pine	..	..	..	..	3 „ 12 „	..
	Nut tree	..	..	..	..	5 „ 7 „	..
	Elder	..	..	..	..	11 „ 9 „	..

* These calculations are based upon the assumption, that the proportion of leaves to branch-wood would be the same as was found to exist on those collected for our local investigation.

† The trunk-wood of the E. obliqua was from an old tree, and scarcely so healthy as could have been desired; this may account to some extent for the small pore stage of potash which it was found to contain.

‡ These numbers correspond very closely with the mean of results obtained by Saussure, Vauquelin, Pertuis, and Kirwan.

Melbourne price of pearlash 45s. to 56s. per cwt.

CONTENTS OF IODINE IN THE LARGE LEATHERY NATIVE SEAWEED, URVILLEA POTATORUM.

Percentage of ash in weed dried at 212° F., 26·51. Percentage of Iodine in the ash, 0·353 (mean of two analyses). The ash likewise contains small quantities of bromine, the amount of which, however, was not estimated.

This plant, when air-dried, measured 13 feet in length, breadth of main leaf 6 inches, thickness of leaf one-sixteenth inch.

TABLE SHOWING THE PERCENTAGE OF ASH, AND OF IODINE IN THE ASH, OF CERTAIN EUROPEAN ALGAE.

Species of Seaweed.	Locality.	Percentage of Ash in Weed dried at 212° F.	Percentage of Iodine in Ash.	Authority.
Laminaria saccharina ..	North Sea	9·78	3·98	Schweitzer.
Fucus digitatus	Mouth of the Clyde	20·49	3·06	Göckobens.
serratus	Mouth of the Clyde	15·63	1·13	Gödechens.
serratus	North Sea	25·83	0·23	Schweitzer.
nodosus	Mouth of the Clyde	16·19	0·46	Gödechens.
vesiculosus	Mouth of the Clyde	18·39	0·31	Gödechens.
vesiculosus	North Sea	20·56	0·11	Schweitzer.

SUPPLEMENTAL LIST OF WORKS CONTAINED IN THE BOTANICAL LIBRARY.

Acta Universit. Lundens. 1864 and 1865.
Agardh, Recensio Pteridis.
Icones Algarum.
de Cellula Vegetabili.
Amtlicher Bericht der Aerzte und Naturforscher Versammlung zu Mainz, 1842.
Anderson, Ceylon Acanthaceæ.
Annales de l'Institut d'Afrique.
Ascherson, ueber Schweinefurthia.
ueber Anticharis.
Atti della Societa d'Acclimatione e di Agricultura in Sicilia.
Baillon, Adansonia. Vols. 3-6.
La Symétrie et l'Organogénie Florale des Marantées.
Description du Genre Longetia.
Etudes sur l'Herbier du Gabou.
Du Genro Netton.
Notice sur les Travaux Scientifiques.
Recherches sur l'Organisation des Caprifoliacées.
Sur le Fruit des Morées.
Fleures Femelles des Couifères.
De la Famille des Aurantiacées.
Bary, Getraide Rost.
Puccinia.
Champignons Parasitiques.
Zoospores.
Saprolegium.
Untersuchungen ueber Uredineen.
Sur la Formation des Zoospores.
Le Développement des Champignons Parasites.
Bayer, Pflanzenformen.
Bentham and Mueller, Flor. Austral. Vol. II.-IV.
Beddome, Ferns of Brit. India. Fasc. I.-XX.
Bericht ueber Versammlung des Hamburg. Vereins.
Bery, Index Libror. Botan. Biblioth. II. B. Petropolit.
Bommer, Monographie de la Classe des Fougères.
Boston, Condit. of Soc. of Nat. History.
Botanische Zeitung. Mohl und Schl. (Fragm.)
Brady, Ailanthus Silkworm.
Braun, Revision des Genus Najas.
Kenntniss von Selaginella.
Bromfield, Plants in the Isle of Wight.
Brown, R., Miscell. Botanical Works. Vol. 2.
Brown, R., jun., on the Nature of Discoloration of the Arctic Sea.
Brown, J., Report of Colon. Botanist, Cape of Good Hope, 1860-64.
Buchenau, Eight Miscell. Writings.
Bulletin de la Soc. Botanique de France, 1854-65.
de la Soc. Impériale d'Acclimation, Paris, 1865-68.
de la Soc. Impér. des Naturalistes de Moscou, 1866-68.
de la Soc. Impér. d'Agriculture d'Algérie.
Bureau, la Famille des Loganiacées.
Sur le Genre Regesia.
Genre Nouveau Schizopsis.
Herborisations de Nantes.
Bignoniacées de la Nouv. Calédonie.
Fleurs Monstrueuses de Primula Sinensis.
Fleurs Monstrueuses de Streptocarpus Rhexii.
Lettre de M. Léon Bureau.
Monogr. des Bignoniacées.
Observat. de la Basse Loire.
Révis. des genres Tynauthus et Lundia.
Caruel, Florula di Monte Christo, 1864.
Catalogue of the Library of Linnean Soc., London.
Parts I. and II.
Clarke on the Natural System of Botany.
Cleghorn, Mem. on Supply of Wood Fuel to the Punjaub.
Colenso, Essay on Botany of New Zealaud.
Collinson, Hortus Collinsonianus.
Cooke, Hardwicke's Science Gossip.
Cosson, Exploration de l'Algérie.
Caudolle, Dissertatio de Memecyleis.
Prodrom. Vol. 15-16.
Note sur un Nouveau Caractère dans les Chênes.
Darwin, the Variations of Animals and Plants under Domestication. 2 vols.
Dillwyn, Fauna and Flora of Swansea.
Drejer, Flora Excursoria Hafniensis.
Falco, Chem. Untersuchung der Rinde von Petalostigma.
Flourens, l'Origine des Espèces.

Fraas, Giftwiesen, N. America und West Austral.
Francis, Lecture on Rust.
Grœlls, Plantas Espanolas.
Indicatio Plantarum Novarum.
Gray, Manual of Botany. 4th and 5th edit.
Grenier et Gordon, Flore de France.
Hance, Notæ Stirpium Asim Orient.
Hannaford, Wild Flowers of Tasmania.
Hanstein, Milchsaft Gefæsse.
Richtungen der Neueren Pflanzen Physiologie.
Rede ueber Pflanzeu Physiologie.
Ueber Befruchtung und Entwickelung von Marsilea.
Harvey, New Algæ of Japan, &c.
Hegelmaier, Mouographie der Gattung Callitriche.
Systematik ueber Callitriche.
Heiberg, Botanisk Tideskrift.
Helms, Die Kartoffel Krankheit.
Der Obstbau an der Nord und Ost See.
Henkel und Hochstetter, Synopsis der Nadelhœlzer Deutschlands.
Herder, Enumeratio Plantar. cis- et transilient.
Periodische Entwickelung d. Pflanzen.
Henry, on Pure Hybridisation and Crossing.
Hildebrandt, Ueber Trimorphismus der Bluethen in der Gattung Oxalis.
Hill, Reports on Bot. Gard., Brisbane.
Hogg, Fruit Manual.
Hooker, Bot. Magazine, 1866-1868.
Synopsis Filicum.
Report on Kew Garden, 1867.
Flora Boreali-Americana.
The Article Botany.
and Arnott, Flora of South America and Pacific Islands.
J., Flor. Tasman., colored copy.
On Insular Floras.
Handbook of the New Zeal. Flora. Vol. II.
Index Semin Hort. Berolin, 1867.
Jaubert, Sur l'Enseignement de la Botanique.
Sur la Végétation du Centre de la France.
Sur Quelques Plantes du Haut Pérou.
Une Lacune dans les Institutions Botaniques.
Sur le Euphorbiacées.
Jardins de Naples.
Johnson, Culture of the Rose.
Journal of the Society of Arts.
Karsten, Plantæ Columbianæ.
Keruer, Gute und Schlechte Arten.
Kirchenpauer, Neue Sertularien.
Klatt, Monogr. von Lysimachia.
Knight, Experim. on the Fœcundation of Vegetables.
Kotschy, l'urisse von Sued-Palestina's Fruchlings-Flora.
Lange, Index Semin., 1861-1866, with notes.
Leighton, Glands on the Phyllod. of Acacia Magnifica.
Lindley, Gardeners' Chronicles, 1866-1868.
Lindley and Moore, The Treasury of Botany. 2 vols.
Lorentz, Biologie and Geograph. der Laubmoose, 1860.
Macedo, Notice sur le Palmier Carnauba.
Mac Owan, Catalogue of South Africau Plants.
Malherbe, de Chêne Liège.
Mann, Hawayan Plants.
Mark, Report on Atoeln Plant.
Martius, Flora Brasiliensis: Dilleniaceæ, Sapotaceæ, Primulaceæ, Myrsineæ, Ericauloneæ, Capparideæ, Fumariaceæ, Scrophularineæ, Cruciferæ, Papaveraceæ, Coniferæ, Cyendem, Barringtoniaceæ, Monocotyledoneæ partim.
Historia Palmarum.
De Sœmmeringia.
Agarici.
Commentar ueber Maregrav und Piso.
Palmeu der alten Welt.
Vernischte Schriften.
Kœnigl. Herbarium zu Muenchen.
Labatin et Pouteria.
Herb. Flora Brazil.
Imperia Flora.
Schriften Botanischen Inhalts.
Versuch eines Commentars ueber die Pflanzen Brasiliens (I. Cryptog).
Beitraz zur Flora Brasiliensis.
Specim. Materiæ Medicæ Brasiliensis.

Maximilian, Prinz von Neuwied, Beitræge zur Flora Brasil.
Metsch, Rubi Hennebergenses.
Mettenius, Filices.
Meyer, Alyssum.
Miquel, Journal de Botanique Néerlandaise.
De Palmis Archipel. Indie.
Annales Mus. Lugdun. 1865-67.
Piperaceœ Flor. Nor. Holl.
Morren et Decaisne, Observat. sur la Flore de Japon.
Mueller, F., Notes sur la Végétation de l'Australie.
Fragm. Phyt. Austr., XXXIV.-XLVIII.
Muenchen Gelehnter Anzeiger, 1855-1860.
Nachrichten der Kœnigl. Gesellschaft der Wissenschaften, Gœttingen.
Nachrichten der Gœttinger Universitæt.
Naturkundige Tijdschrift voor Nederl. Indie, 1866.
Naudin, Observat. Plantes Hybrides.
Newton, Report of the Department of Agriculture, United States, 1864.
Nyman, Sylloge Floræ Europeœ. 2 vols.
Oliver, Guide to the Royal Bot. Gardens, Kew, 1865 and 1866.
Paine, Catalogue of Plants, Oneida Country, 1865.
Pancic, Plantæ Serbicœ, I. and II.
Pappe, Botanical Reports.
Parlatore, Flora Palermitana.
Plantæ Novæ.
Payer, Leçons sur les Familles Naturelles. 6 fasc.
Elemeuts Botaniques.
Notice sur les Travaux Scientifiques.
Organogénie. 3 fasc.
De la Famille des Malvacées.
Report sur les Mémoires.
Planchon, Hortus Donatiensis.
Pœch, Plant. Insul. Cypri.
Proceedings of Linn. Soc., 1866-68.
Pritzel, Iconum Botan. Index, 2nd edit.
„ Supplement.
Pursch, Flora Americana. 2 vols.
Rapport sur l'Exposition Hortic. de Nantes.
Record of the Intercolonial Exhibition.
Regel, Die Gærten von Petersburg.
Sur le Valeur de l'Espèce.
Index Semin., 1866.
Russische Aepfelsorten.
Reports of the Regents of Univers. of New York, 18th and 19th.
Reports of Proceedings of International Horticult. Exhibit.
Revue des Cours Scientifiques. 2 fasc.
Reynosa, Ensays s. e. Cultivo de la Canna de Azucar.
Robin, Histoire des Végétaux Parasites.
Rossman, Beitrœge zur Kentniss der Wasserhahnenfuessœ.

Royle, Fibrous Plants of India.
Salm-Dyk, Monographie von Aloe und Mesembryanth. 6 volum.
Schlechtendal, Hortus Halensis. 2 fasc.
Plantæ Wagnerianœ.
Schomburgk, Report on Bot. Gard. of Adelaide.
Schrank, Primitiœ Florœ Salisburg.
Schrenk, Enumeratio Plant. Nov.
„ Nov. Petrop.
Schumacher, Enum. Plantarum Sællandiœ.
Schultz, Ueber die Tanacetœn.
Seemann, Flora Vitiensis. Part II.-VI.
Journ. of Bot., 1866-1868.
Steinheil, Observat. on Zanniehellia.
Suringar, Observationes Phycologicæ in Floram Batavam.
Schnitzlein, Iconographia, 1866-67.
Tassi, Notes on Hoya. Magnolia, and Aquilegia.
La Vita Dei Flori.
Sulla Flora Toscana.
Teysmann, Capellinia.
and Binnendijk, Hort. Bogoriensis.
Thozet, Notes on Roots. Fruits, &c., in use as food by Aboriginals in Queensland.
Thwaites, Report of Royal Bot. Garden, Peradenia, 1866-1867.
Tinne et Kotschy, Plantæ Tinneanæ.
Todaro, Synops. Acotyl. Vascul.
Torrey and Gray, Report Bot. of N. Amer. Expedition.
Transact. Linnean Society, 1866-68.
Treviranus, De Nymphæa et Euryale.
Die Mistel.
Trevisan, De Dictyonis Adumbratio.
Brigantiæa, Novum Lichenum Genus.
Trottier, Note sur l'Eucalyptus et sur le Reboisement de l'Algérie.
Tulasne Discurs du Select. Fung. Carpolog.
Uebersicht der Thœtigkeit des Naturwissen schaftlichen. Vereins Hamburg.
Verhandlungen des Botan. Vereins fuer Provinz Brandenburg, 1859-65.
Verzeichniss der Sammlungen zu Mainz. L. Abthlg. 1843.
Vieillard, Etudes sur Oxera et Deplauchea.
Plantes Utiles de la Nouvelle Calédonie, 1865.
Plaut. Interessant. Nouv. Caléd.
Filices Novæ Caledoniœ.
Vilmorin et Andrieux. Les Fleurs de Pleine Terre.
Visiani. Plant. Serbicar. Pemptas.
Plantæ Fossili.
Walper's Annal. Bot. Syst. Tom. VII.
Woolls' Contributions to the Flora of Austral.
Wittmark on Musa Eusote.
Wittstein, Etymol. Bot. Wœrterbuch.
Zollinger, Auonaceœ des Ost Ind. Archipel.

CONTRIBUTIONS OF GROWING PLANTS RECEIVED SINCE LAST REPORT.

Acclimation Society, Brisbane.
„ „ Dunedin.
„ „ Melbourne.
„ „ Otago.
„ „ Paris.
Addis, J., Collingwood.
Allport, M., Hobarton.
Altmann, Mrs., Richmond.
Amstel, P. Van, Melbourne.
Bailey, S. W., Belfast.
Barkly, Sir H., Mauritius.
Bernays, L., Brisbane.
Botanical Garden, Adelaide.
„ „ Bourbon.
„ „ Brisbane.
„ „ Calcutta.
„ „ Cape of Good Hope.
„ „ Ceylon.
„ „ Edinburgh.
„ „ Geelong.
„ „ Hobarton.
„ „ Hong Kong.
„ „ Kew.
„ „ Liverpool.

Botanical Garden, Mauritius.
„ „ New Caledonia.
„ „ Portland.
„ „ Sydney.
„ „ Trinidad.
Boyd, Capt., Vicksburg.
Brewer, T., New Zealand.
Bright, C. and Reg., Melbourne.
Bromby, Dr., South Yarra.
Brown, Rev. Dr., Capetown.
Brunning, G., St. Kilda.
Butler, T. H., Calcutta.
Caldwell, S., Escape Cliffs.
Callender, M., Prahran.
Cameron, D., South Yarra.
Campbell, D. S., Richmond.
Carter, W., Emerald Hill.
Cashmore, D., Geelong.
Castella, H. de, Yering.
Claverhill, J. C., Hawswood, New Zealand.
Chamberlain, C. F., Ballarat.
Celay, Chas., Sandridge.
Clifton, W., Albany.
Clough, J. H., Melbourne.

No. 21, b.

Cole, T. C., Boroondara.
Cole, T. J., Richmond.
Dardell, J., Batesford.
Davis, Archdeacon, Hobarton.
Duerdin, J., Melbourne.
Dynes, H., Richmond.
Ellery, Dr., Melbourne.
Fallshaw, D., Cheltenham.
Fawcett, J., Richmond River.
Fearney, Miss, Malmsbury.
Forwood, W. H., Melbourne.
Frazer, Miss F., ship *Dayspring*.
French, C., South Yarra.
Gaskel, J., Prahran.
Gelding. J. W., Sydney.
Glass. H., Flemington.
Gleadow. R., Gippsland.
Gibson, J., Melbourne.
Goulding, G., Melbourne Cemetery.
Grosse. F., Melbourne.
Guilfoyle and Son, Sydney.
Gulliver, B., Melbourne.
Handasyde, J., Melbourne.
Hannaford, S., Launceston.
Harding. J., Hawke's Bay.
Harris, J., South Yarra.
Helms, C., Hokitiki.
Henderson, J., Richmond River.
Hendersen, J., Sydney.
Hirschi, J., Castlemaine.
Holroyd, A., Sydney.
Horticultural Society, Melbourne.
Howell, M. C., Prahran.
Howitt, Dr., Melbourne.
Iron, J., Echuca.
Jones, P. J., Melbourne.
Joshua Brothers, Melbourne.
Kelleway, W., Wellington.
Kendall, Mrs., South Yarra.
Kyte, A., Melbourne.
Lamont, W., Hawthorn.
Lang and Co., Ballarat.
Law, W., Northcote.
Lee, W., Collingwood.
Lenne, C., Castlemaine.
Lewis, Panama Mail Company.
Liffen, J., Greymouth, New Zealand.
Ligar, C., Surveyor-General.
Lightfoot, T., South Yarra.
Lipscomb, W., Hobarton.
Macarthur, Sir Will., Camden.
Maclean, Capt. of *Alexandra*.
Macmillan, T., Melbourne.
MacSorby, Rev., Albany, West Australia.
Manifold, Mrs., South Yarra.
Manners-Sutton, His Excellency and Lady.
Marsden, Mrs., St. Kilda.
Maxwell, G., Albany, West Australia.

Moffat, G., Prahran.
Mont, S., Captain of *Elizabeth*.
Morrell, J., Melbourne.
Moyle, J., Prahran.
Murray, A., Melbourne.
Noone, J., Melbourne.
Officer, Dr., Hobarton.
O'Shancey, P., Rockhampton.
Oatler, J., Toongabbie.
Parker, J. H., Hokitiki.
Perry, Bishop, Melbourne.
Perry Brothers, Melbourne.
Phelan, T., Prahran.
Phillips, P., Hawthorn.
Ralston, A. J., Sydney.
Ramel. P., Paris.
Randall, Capt. of ship *Otago*.
Ridge, J., Prahran.
Robertson, Dr., Queenscliff.
Ronalds, N., Richmond.
Rusden, W., Brighton.
Sangster and Taylor, Toorak.
Shand, Christchurch, N.Z.
Sharp, T., Collingwood.
Shephord and Co., Sydney.
Scott, J., Hawthorn.
Scott, J. T., Lancefield.
Seidel, B., Geelong.
Smith, L. L., Melbourne.
Smith and Sons, Riddell's Creek.
Sonder, Dr., Hamburg.
Stone, C., Brighton.
Stork. J., Fiji Islands.
Stuart, C., Timbarra.
Sydney, J., Melbourne.
Synnott, M., Terricks.
Thomson, Mrs. A., St. Kilda Road.
Thozet, A., Rockhampton.
Tulk, A., Melbourne.
Turner, Mrs., South Yarra.
Tyers, C., Port Albert.
Tyesmann. J., Java.
Ulph, F. B., Otago.
Verschaffelt. A., Ghent.
Virgoe, J., Brighton.
Walsh, G., South Yarra.
Watson, W., South Yarra.
Watts, J., Richmond.
Weatheral, C., Cheltenham.
Webster, G., Richmond.
Were, J. B., Melbourne.
Westwood, J., River Forth, Tasmania.
Whatmough, R., Greensborough.
Wilhelmi, C., Prahran.
Wills, Miss K., Richmond.
Winterstein, A., Alexandria.
Young, J., Sydney.

CONTRIBUTION OF SEEDS.

Abbott, T., Hobarton.
Acclimation Society, Canterbury.
    „   „   Dunedin.
    „   „   Melbourne.
Agrer, Don. Frontera.
Agricultural, Board of, Melbourne.
Agricultural Department, United States America.
Agri-Horticultural Society, Lahore.
Aitken. D. W., Singapore.
Aldis, W. H., Melbourne.
Alexander, A., Melbourne.
Allanby, J., St. Arnaud.
Allport, M., Hobarton.
Amery, C., Lahore.
Amstel, Ploos Van, Melbourne.
Atkinson, Miss, Berrima, New South Wales.
Barley, Hon., Col. Sec., W. Australia.
Balfour, Prof., Edinburgh.
Barkly, Sir Henry, Mauritius.
Barry, Sir Redmond, Melbourne.
Bates, C., Melbourne.

Beal, O., Melbourne.
Beames, W., San Francisco.
Beattie, H., Albury.
Beddome, Major, Madras.
Bell, J. A., Richmond.
Benary, E., Erfurt.
Bennett, Dr., Sydney.
Beverly. A., Dunedin.
Bicknell. F., Dunedin.
Bissill, W., Ravenswood.
Bjornen, F., Fitzroy.
Blackburn. Captain. Williamstown.
Blair, J., Melbourne.
Bland, C. J., Clunes.
Bolander, Dr., San Francisco.
Bond, R., London.
Bonney, F., Mount Murchison.
Bostock, Rev. J., Fremantle.
Botanical Garden, Adelaide.
    „   „   Bonn.
    „   „   Brisbane.
    „   „   Buitenzorg.

Botanical Garden, Calcutta.
" " Cape of Good Hope.
" " Ceylon.
" " Copenhagen.
" " Florence.
" " Geelong.
" " Grahamstown.
" " Kew.
" " Liverpool.
" " Lyon.
" " Madrid.
" " Montpellier.
" " Natal.
" " New Caledonia.
" " Palermo.
" " Parma.
" " Petersburg.
" " Pisa.
" " Saharunpoor.
" " Siena.
" " Sydney.
" " Turin.
" " Vienna.
Brewster, R., Penola.
Brown, Dr., Cape of Good Hope.
Bruggen, Captain of *Friendship*.
Bull, W. B., Castlemaine.
Bull, W., Chelsea.
Bull, —, London.
Butler, T. H., Calcutta.
Butler, J., Sandhurst.
Camaldis, A., Melbourne.
Camara, A., Sydney.
Campbell, D. M., Avon Downs, Queensland.
Carey, C., Dunbury, Western Australia.
Carige, Mrs., Benalla.
Cartwell, J., Sandhurst.
Cleghorn, Dr., India.
Clifford, G. P., Dunedin.
Clough, C. H., Melbourne.
Cobham, Mrs., Sydney.
Cole, Commiss. Euston.
Conn, W. G., Bowen.
Cooper, Rev. W. C.
Curr, E., Melbourne.
Damyon, F., Melbourne.
Darbyshire, J., Sutton Grange.
Davis, P., Prahran.
Denison, Sir W., Madras.
Denny, Dr., Inglewood.
Diaper, J., Calcutta.
Dinsdale, Captain.
Dobbyn, W. A., Wangaratta.
Dobson, F., Melbourne.
Donaldson, D., Hong Kong.
Dowling, Dr., of *Swiftsure*.
Draeger, C., Melbourne.
Drysdale, A., Chief Officer of *Western*.
Duncan, W., Prahran.
Dunn, E., Benalla.
Edmondson, R., Melbourne.
Edwards, H., San Francisco.
Edwards, J. N., Brisbane.
Evans, Dr., New Zealand.
Fawcett, C., Richmond River.
Featherstone, Dr., Wellington.
Ferguson, F., Camden.
Fletcher, G., Sandhurst.
Forsyth, Mrs. E., Port Chalmers.
Gibraltar, Colonial Secretary.
Gill, E., St. Kilda.
Gould, Dr., New Zealand.
Goulding, W., Melbourne Cemetery.
Grainger, Captain of *Rangatira*.
Gryne, Captain of *Eliza*.
Guilfoyle and Son, Sydney.
Gray, A., Prof., Boston.
Greeves, Dr., Melbourne.
Grey, Sir G., New Zealand.
Haage and Schmidt, Erfurt.
Hall, J., Hastings.
Handasyde and MacMillan, Melbourne.
Hannaford, S., Launceston.
Harding, J., Hawke's Bay, New Zealand.
Hardy, J., Emerald Hill.
Harris, J., South Yarra.
Head, W., Canterbury, New Zealand.
Heathcote, J. W., Fowler's Bay.

Hector, Dr., Canterbury.
Heineman, C. G., Melbourne.
Helicar, W., Melbourne.
Helicar, J., Melbourne.
Helms, B., Melbourne.
Henderson, Dr., Lahore.
Henderson, Chief Constable, Richmond River.
Hetherington, Rev. J., Melbourne.
Higgins, F., Elsternwick.
Hockins, A. J., Brisbane.
Holdsworth, J., Melbourne.
Hood, J., Melbourne.
Hooley, J., Nichol Bay.
Howard, J., Beechworth.
Huber frères, Hyères, France.
Hughan, A., Melbourne.
Hughes, C., Melbourne.
Hyndman, W., Melbourne.
Ick, E., Melbourne.
Inglis, E., Emerald Hill.
Jaffray, W., Melbourne.
Jamison, Dr., Saharunpoor.
Jefferson, Mrs., Fernshaw.
Kean, J. B., Dunedin.
Kelleway, W., Wellington.
Kendall, F. R., Melbourne.
Kidney, Captain of Steamer *Albion*.
King, G., Melbourne.
Kirk, T., Auckland.
Kraemer, F., Sandhurst.
Labertouche, Mrs., Melbourne.
Lamb, D., Richmond.
Lane, C., Queenscliff.
Langton, H., Richmond.
Law, Somner, and Co., Melbourne.
Lawrence, J., Melbourne.
Layard, Cape of Good Hope.
Lenehan, Rev. T., Melbourne.
Lenormaud, R., Vire, France.
Lewis, T., Omeo.
Lindsay, Mrs., Richmond.
Lloyd, E., Melbourne.
Macarthur, Sir Will., Camden.
MacCrae, D., Melbourne.
McCoy, Profess.
MacKinlay, Adelaide.
MacLeod, J., Melbourne.
MacMillan and Grant, Melbourne.
MacSorby, Albany, Western Australia.
Madden, Dr., Melbourne.
Manners Sutton, His Excell., and Lady.
Manzie, D., Richmond.
Mappin, W., Sandhurst.
Martichon, Pere et fils, Cannes.
Martelli, A., Melbourne.
Mason, T., Wellington.
Matson, J. M., Melbourne.
Maxwell, Albany.
Mercer, W., West Prov., India.
Michael, Major, Madras.
Miersch, W., Benalla.
Moody, J., St. Kilda.
Mooney, Miss O., St. Kilda.
Moore, Dr., Adelaide.
Moore, D., Melbourne.
Moore, S. W., San Francisco.
Morton, L., Melbourne.
Murray, A., Hong Kong.
Napier, J., Moonee Ponds.
Naturforscher, Verein, Bremen.
Negus, P., Melbourne.
Nicholson, G., Melbourne.
Nicholson, N., Melbourne.
Noone, J., Melbourne.
Norman, Captain of H.M.C.S. *Victoria*.
Opitz, F., Melbourne.
Osborne, J., Boston.
O'Shanesy, P., Rockhampton.
Oswald and Inglis, Melbourne.
Paris, Mus. d'Hist. Natur.
Parker, A. M., of *Swiftsure*.
Paterson, J. M., Meredith.
Pentzke, Th., Spring Creek, Queensland.
Perry, Bishop, Melbourne.
Perry, C. J., Fitzroy.
Philipi, Prof., Santiago.
Playford, C. R., Stawell.
Politz, J., Richmond.

Poore, Rev. T., St. Kilda.
Purdie, Dr., Dunedin.
Pury, S. de, Yering.
Ralston, A., Sydney.
Ramel, P., Paris.
Reynolds, J. N., Melbourne.
Riddel, J. C., Mount Macedon.
Ridgell, E., Woodspoint.
Robinson, G., Berwick.
Robinson, Sir H., Ceylon.
Ronalds. J. S., Sandhurst.
Rothwell, E., Sandhurst.
Rose, J. H., South Yarra.
Ross, J., Melbourne.
Russel, R., Melbourne.
Russel, J., Grant.
Sayce and Co., Melbourne.
Schlotthuber, Dr., Goettingen.
Shaw, H., Melbourne.
Smith, Rev. J., Delhi.
Smith, J. T., Melbourne.
Smith, J., Mansfield.
Smith and Sons, Riddell's Creek.
Smith and Symons, Glasgow.
Sonder, Dr., Hamburg.
Soues, E., Melbourne.
Spruce, J. D., Assam.
Sterndale, Captain, Melbourne.
Stone, J., Glenpatrick.
Stuart, C., Timbarra.
Sturm, C., Napier, New Zealand.
Sumner, T., Melbourne.
Sweeney, W. and Co., San Francisco.

Syder, D., Melbourne.
Synnott, M., Terricks.
Taylor, J., Richmond.
Thatcher, C. E., Fitzroy.
Thomas, M., Collingwood.
Thozet, A., Rockhampton.
Tiffon, H., Napier, Now Zealand.
Tripp, W. R., Sydney.
Tripp, Miss F., Prahran.
Turnbull, A. B., Melbourne.
Van Delden, Java.
Vilmorin, Andrieux, and Company, Paris.
Vivian, M., Maldon.
Wakefield, F., Wellington.
Walker, W. C., San Francisco.
Walker, Miss, Launceston.
Wallcott. P., Karridale.
Walter, C., Lilydale.
Wehl, Dr., Mount Gambier.
Wharton, Miss, Glasgow.
Wilhelmi, C., Prahran.
Wilkins, Mrs. A., St. Kilda.
Wilkinson, H., St. Kilda.
Wilmore, T., Williamstown.
Wilson, Mrs., Jolimont.
Wilson, S., Wimmera.
Winterstein, A., Alexandria.
Wood, J., Sandhurst.
Woolridge, Dr., South Yarra.
Woolls, W., Paramatta.
Yarwood, H., Round Hill Station, New South Wales.

CONTRIBUTIONS OF MUSEUM PLANTS.

Agardh, Prof., Lund.
Allen, J. M., Sale.
Anderson, Dr., Calcutta.
Atkinson, Miss, Berrima.
Babbage, H., Adelaide.
Baillou, Prof., Paris.
Baudinet, E., Kent's Group.
Baudinet, Miss, Swan Island.
Beddome. Major, Madras.
Beckett, W. N., Ceylon.
Beverly, J., Dunedin.
Bissill, W., Ravenswood.
Bloxam, A. R., New Zealand.
Bonney, F., Mount Murchison.
Bowman, E., Rockhampton.
Boyle, D., Nunawading.
Brubon, J., Swan Hill.
Bucknall, H., Natal.
Bull, W., Tumberumba.
Bureau, Dr., Paris.
Canby, W. M., Boston.
Carron, W., Sydney.
Cosson, Dr., Paris.
Dalton, J. C., Curriwillighi.
Decaisne, Prof., Paris.
Dettmar, C., Dallinga.
Dickie, Prof., Aberdeen.
Diggles, S., Brisbane.
Faric, Dr. R., Loddon.
Fawcett, C., Richmond River.
Fereday, Rev., Georgetown.
Fitzalan, E., Port Denison.
Forde, Mrs., Hunter River.
Fullagher, P., Little River.
Galloway, J., Melbourne.
Glendinning, Mrs. Maldon.
Giles, E., Mount Murchison.
Goodwin, Miss, Launceston.
Graeffe, Dr., Samoa.
Gray, A., Professor, Boston.
Gulliver, B. and T., Melbourne.
Haast, Dr., Christchurch.
Haly, C. A., Queensland.
Hall, J., Hastings.
Hance, Dr., Whampoa.
Harper, C., Port Walcott.
Hannaford, S., Launceston.

Henderson, J., Richmond River.
Hooker, Dr., Kew.
Hughan, A., Melbourne.
Hulis, W., Escape Cliff.
Kelleway, W., Wellington.
Kirk, T., Auckland.
Knight, E., Porongerup.
Kurz, Dr., Calcutta.
Lane, H., Yendillah.
La Trobe, C. J.
Lenormand, R., Vire, France.
Liffen, T. H., New Zealand.
MacGee. Captain of the Schoolboy.
MacKinlay, J., Gawlertown.
MacOwan, P. Grahamstown.
Martin, Dr., Champion Bay.
Maxwell, G., Albany.
Meller, Dr., Mauritius.
Milligan, Dr., London.
Miquel, Prof., Utrecht.
Moore, C., Sydney.
Morton, L., South Yarra.
Mueller, T., Melbourne.
Nernst, J., Port Mackay.
Nott, Mrs., Maldon.
O'Shanesy. P., Rockhampton.
Pancher, J., New Caledonia.
Perceval, de Grandmaison, Paris.
Parlatore. Prof., Florence.
Peechly, Dr., Darling River.
Ramsay, E. P., Sydney.
Regel, Prof., Petersburg.
Robinson, G., Berwick.
Ross, Miss M., Dingy, New South Wales.
Sheridan, R. B., Maryborough.
Story, Dr., Swanport.
Strong, E. N., King's Island.
Stuart, C., Timbarra.
Sturm, C., Napier, New Zealand.
Thozet, A., Rockhampton.
Thwaites, Dr., Ceylon.
Travers, W. F., Christchurch.
Tyesman, J. E., Java.
Walcott, P., Karridale.
Walter, C., Lilydale.
Whan, Rev. W., Skipton.
Woolls, W., Paramatta.

SUPPLEMENTARY LIST OF CONTRIBUTORS OF GROWING PLANTS

RECEIVED FROM SEPTEMBER, 1868, TILL MARCH, 1869.

Many donations have besides been received from donors whose names occur in the foregoing pages.

Balldy, G., Collingwood.
Bernays, A., Brisbane.
Botanic Garden, Grahamstown.
„ „ Brisbane.
„ „ Natal.
Bryerly, T., Brisbane.
Cassel, Mrs., Prahran.
Comrie, Curryong, N.S.W.
Cook, G., Prahran.
Daintree, Richard, Gilbert River.
Ewen, J. A., New Zealand.
Farie, Claud, South Yarra.
Huber Frères, Hyères, France.

Learmonth, A. L., Irsledon.
MacOwen, P., Grahamstown.
Martin, J. P., Richmond.
Meller, Dr., Botanic Garden, Mauritius.
Muir, P., Brisbane.
Ryan, Mrs. Ch., Richmond.
Shepherd and Co., Sydney.
Sheridan, R. B., Maryborough, Queensland.
Sims, J. B., Brighton.
Todaro, Professor, Palermo.
Virgoe, W., Brighton.
Weidenbach, Max., Adelaide.

CONTRIBUTORS OF SEEDS.

Acclimatisation Society, Brisbane.
„ „ Otago.
Agri-Horticultural Society, Punjaub.
Amsinck, Capt., R.N., Melbourne.
Bernays, L. A., Brisbane.
Birchall, H., Richmond.
Bryant, T., New Zealand.
Bruce, J. D., Assam, India.
Carey, J. C., Bunbury, Western Australia.
Chapman, D., Western Australia.
Cresswell, C. A., Hobarton.
Daintree, Richard, Gilbert River.
Evans, R., Collingwood.
Fitzalan, E., Port Denison, Queensland.
Gerrard, J., Sale, Gippsland.
Government, Madras, India.
Graburn, W., Otago.
Graham, W., Fiji.
Hamilton, W. S., Wellington.
Hicks, J., Melbourne.
Hill, W., Brisbane.
Hooker, Dr. J. D., Kew.

Howitt, Dr., Melbourne.
Howitt, A. W., Bairnsdale.
Horticultural Society, Melbourne.
Huddlestone, F., Otago.
Jouret, A., Melbourne.
Katzenstein, J., Cassel, Germany.
Kenworthy, E., Melbourne.
Langtree, H., Richmond.
MacKen, Botanic Garden, Natal.
MacKenzie, Dickson, Assam, India.
Miller, W. H., South Yarra.
Pancher, A., New Caledonia.
Regel, Dr. E., St. Petersburg.
Sangster and Taylor, Toorak.
Shand, C., Christchurch, N.Z.
Sichel, E. F., Melbourne.
Stuart, C., Tenterfield, New England.
Synnott, W., Grant.
Thwaites, Dr., Ceylon.
Walker, W., Melbourne.
Williams, Dr., Queenscliff.
Youl, Dr., Melbourne.

CONTRIBUTORS OF MUSEUM PLANTS.

Barlee, F., Colonial Secretary, Perth.
Beckett, T. W. N., Ceylon.
Bureau, Dr. E., Paris.
Calvert, J. G., Cavan, Yass, Sydney.
Carey, C., Bunbury, W. A.
Chapman, D., Western Australia.
Fitzalan, E., Port Denison, Queensland.
Fowler, W., York's Peninsula.
Gummow, Dr., Swan Hill.

Helmich, A., Perth, W.A.
Jones, D., Auckland, N.Z.
Knight, W. G., Porongerup, W.A.
Mein, Dr., Edwards River.
Musgrave, A. W., Warrnambool.
Sullivan, Dr., Gawler Ranges.
Todaro, Professor, Palermo.
Warburton, G. E., Western Australia.

A series of Victorian geological specimens, particularly illustrative of soils in relation to vegetation, has been deposited in this establishment by the kindness of A. Selwyn, Esq.

A second supplementary list of books and journals secured recently for the library will be furnished hereafter.

By Authority: JOHN FERRES, Government Printer, Melbourne.

www.ingramcontent.com/pod-product-compliance
Lightning Source LLC
Chambersburg PA
CBHW022034190326
41519CB00010B/1713